Introduction

  Everyone has their own personal reason for getting into chicken keeping … handed down in the family, to be self sustaining, to sell eggs or chicks, breeding.

My entrance into chicken keeping was an evolution from a medical issue. In 2010 I was diagnosed with Type 2 Diabetes. I had a terrible time with the side effects of medication so my doctor was willing to work with me to control my glucose with diet. After several years of working on this, treatment I ended up on a plant-based diet where my main source of protein is eggs. Between cooking from scratch for baked goods and breads, and eating eggs themselves, I go though 2 — 3 dozen eggs each week. So I decided that if I eat that many eggs they need to be the best eggs I could find.

Through reading the books my doctor recommended and the plant-based diet research I did, I got into organic vs non-gmo, vegetarian vs vegan vs paleo. And after seeing the eggs available at the various grocery stores in my area, I came to the conclusion I needed to raise my own chickens to get the quality eggs I was after. I want no soy, no corn, and no canola in my diet.

My family has always been animal lovers but we didn't know anyone that raised chickens, we don't come from a family of farmers, so I started researching on the internet. I found a local permaculture group that offered a 2 hr class on raising backyard chickens for a nominal fee. I figured the investment in the class would help me determine if this was something I should pursue. The class confirmed this was something we could handle.

The Flock

To keep harmony among the flock they either need to all be the same breed or a thorough mix of different breeds. I chose a mixed breed flock. At the time of this writing my hens that are of laying age are 2 Welsummer, 2 Wheaten Americauna, 1 unknown mixed breed, 1 Buff Orpington, 1 Speckled Sussex, 1 Silver Laced Wyandotte and 1 Gold Laced Wyandotte. I also have 3 Silverudd Blue (Isbar) juveniles.

A flock is a minimum of 3 birds. Incase 1 dies there will be 2 left to still be a flock. 1 chicken won't do well alone.

Brooder - raising chicks

A brooder box is used to raise chicks. This can be a plastic tub, a water trough, a cage, or a custom built brooder. It needs a lid so if using a plastic tub, either poke holes in the plastic lid or put it aside and just use a piece of hardware cloth cut to size.

They will need heat, water, chick starter feed, chick size feeder and waterer, chick grit, and bedding in the box.

Lay an old towel or paper towels on top of the bedding for the first few days so the chicks learn what food is and where to find it. You don't want them to eat the bedding.

To prevent 'pasty butt', put 2 Tbs of ACV per gallon of water. This can be mixed with chick probiotic powder in the same gallon of water.

Hang an old fashioned feather duster in the brooder so the feathers are 2 inches from the bedding. The chicks will use this to hide under and for warmth like they would a momma hen.

Provide smaller versions of items that the chicks will find in the Coop/Hen House. Start with a small diameter roost bar and gradually increase the size as they grow ending with the same size the adults have. They might not actually roost on it, but they will climb on it, jump on it, and eventually roost on it.

If you have the space, give them a small jungle gym … something to play on where they can jump, fly, and chase each other.

If you use a nipple waterer in the Coop/Hen House for the adults, teach the chicks. You can get the nipple attached to a bottle cap, or make it yourself, and use this on a standard bunny cage bottle. At about 3 weeks they will be able to learn to use the bottle nipple waterer. You only need to get 1 chick to do it and the others will follow. They teach each other.

Temperature is important for chicks. Whether you use a heat lamp or heat table, put the heat source at one end of the brooder so the chicks can get as close or as far away as they need. Week 1 the temperature should be 95 degrees, Week 2 should be 90 degrees, and Week 3 should be 85 degrees, dropping 5 degrees each week until 70 degrees. From Week 6,

 for the first 3 months, the minimum temperature should be 70 degrees during the day and 60 degrees at night.

When using a heat lamp, I don't recommend the typical brooder lamp that most livestock stores sell. Instead I prefer a heat lamp that doesn't provide any light. If the chicks are raised with always having light, even low light, they have a harder time transitioning to dark nights later. You can use the brooder lamp shade, but go to a pet store and get a reptile heat bulb. This will emit heat without light.

Once the chicks are about 6 weeks old and fully feathered, they can go outside if they were raised inside, but still adhere to the above temperatures.

As much as you can, give the chicks everything in the brooder that they will have in the Coop/Hen House. This will make transition to the Coop easier for them.

Extreme Heat

If I get newly hatched chicks in the autumn, I bring the brooder inside to raise them. The heat lamp can keep up during the day, but the nights outside can get too cold.

If I get them in the spring or summer, I raise them outside because the heat lamp is only needed at night.

In either case, I put the heat lamp on a timer so I can control the hours that its turned on.

Introducing Chicks or New Members to the Flock

Chicks

Once the chicks have their full feathers, they can move from the brooder to the grow out pen. You don't want to introduce chicks to the adult flock until they are close to the same size as the adults unless they were raised by a flock member momma. The flock member momma will protect them and teach them how to assimilate into the flock.

It is best if the grow out pen shares a hardware cloth wall with the larger flock so they can get acquainted as the chicks grow over the next couple of months.

New Adult Members

If you got new adult members to add to the flock, it is recommended to quarantine them for 30 days to be sure they don't introduce anything unwanted to your flock like infections, parasites, or other health issues.

Final Introduction Step

Adding new members to the flock is stressful and upsets the current pecking order. The flock's instinct is to drive the new members away. The pecking order will need to be re-established with the new members included.

Whether new adults or grown chicks, create an Introduction Cage containing food, water, a roost bar and if laying age a nest box. This can be a section of your run separated by hardware cloth or a metal dog crate placed in the run. The idea is for the new members and current flock to get to know each other without being able to get to each other. The new members will need to stay in this cage for 1-2 weeks.

If using a metal dog cage and they are trying to fight through the wire cage, cover the walls with hardware cloth for the first week. This will allow them to still meet each other but not get to each other. After a week, take the hardware cloth off and there shouldn't be much of a problem.

After a week or two and you're ready to let the new members into the flock, choose a time when you have several days (at least over a weekend) to watch them closely. If the flock doesn't have enough to keep busy they will chase and pick on the new members. So the best thing is to create healthy distractions. Some ideas are hanging a head of cabbage for them to play with, add a place for the new members to hide like under some fresh branches, put food treats in multiple places so the new members don't have to compete and the current flock is busy, have multiple feeders and waterers, rearrange the coop so there is more new than the new members, or if you free range let the new members out on the range where they have space to mingle at their own pace.

Expect a certain amount of bullying until the new pecking order is established. As long as they aren't drawing blood, just watch from the sidelines ready to come to the rescue if needed. If there's too much fighting or if blood is drawn put them back in the Introduction Cage for a few more days and then try again.

The Coop/Hen House

You can buy a coop kit or custom build a coop. Either way there are some basic details to take into consideration.

Use hardware cloth, not chicken wire. Chicken wire isn't strong enough to keep predators out. 1/4 inch hardware cloth for chicks and 1/2 inch hardware cloth for adults.

Not only cover the sides of the run with hardware cloth, but place it 2 inches under the surface of the ground and out another 18 inches from the wall. This is to deter any predators that think they can dig under the wall to get inside to coop.

Each adult bird needs 4 SQFT of coop space and 8-12 SQFT of run space. This will give the flock enough room to move around and get along. Less than this might cause them to pick on each other.

When the flock is new, if the coop is large enough, close them in the coop for 2 weeks. This will associate them with their new home and where they should sleep. When adding new flock members they will follow the lead of the existing flock and not need to be closed in for several days.

Once you've decided how many chickens will be in your flock, multiply it a few times. Chickens are addictive and you will always find reasons to add more … this is called Chicken Math. So be sure to build your coop for the size flock you will have several years down the road.

Extreme Heat

Living in the Arizona desert, out in a county island, we have coyotes, javelina, bobcats, and other animals roaming wild. So we decided we needed a fully enclosed run with hardware cloth walls and a solid roof attached to our coop to keep the chickens safe because we work off the property and can be gone for 10 hours during the day. This also provides as much shade as possible during the day.

Another thing to consider is the paint color for the outside of the coop and run. Darker colors will absorb more heat from the sun than lighter colors.

Nesting Boxes

Hens like to lay in dark places. You can build or buy actual nest boxes, use plastic tubs, or use wooden crates or boxes. Hanging curtains on the front is one way to make it darker inside. To help the hens know where you want them to lay, you can buy fake ceramic eggs to place in the nest boxes.

There are several types of nest boxes. Standard boxes have a flat bottom and are usually built into the wall of the coop with a hatch door to retrieve the eggs from the outside. There are also what is called Rollaway boxes that have a slanted bottom for the eggs to roll away from the laying hen to a collection area either in the front or the back of the box.

You'll need to provide 1 nest box for every 3-4 hens. They probably won't use them all. In fact, they'll probably all try to use the same nest box.

Nest boxes need to be a 12-14 inch cube with a 4-6 inch lip on the front to hold in the bedding.

Extreme Heat

With extreme summer temperatures as high as 118 degrees, you don't want the nest boxes to get too warm from a lot of bedding. I chose to use nest box pads instead. They are made of plastic, washable and provide enough of a cushion that the eggs don't crack when laid.

I also didn't include the divider wall between the boxes. This will stop the hens from fighting over a single box, they can lay together, and it allows more air flow so they don't overheat.

Roost Bars

Some chickens prefer round roost bars and others prefer flat ones. For round roost bars, a closet dowel or tree branch works great. For flat roost bars, a 2x4 wood stud cut to length. Regardless of the type of roost bar, allow at least 1 FT per adult chicken of roosting space.

Roost bars need to be mounted higher than the nest boxes to keep the hens from roosting in the nest boxes. They poop most while sleeping so roosting in the nest box will make the box dirty, giving dirty eggs. Also plan for how to clean the poop under the roost bars so there's enough room for you to comfortably get under there.

Extreme Heat

Even summer nighttime temps are high with a nightly low of 95 degrees and might not get down to a comfortable level. We installed a small barn fan hanging from the ceiling above the roost bars to blow the collecting hot air out a high ventilation window.

Bedding / Litter

Most importantly, the bedding material chosen should not cause respiratory issues, should not compact down and should be absorbent for as long as possible.

On the coop floor the bedding will provide a soft walking surface and absorb poop and it's odor. The bedding should be at least 2 inches thick. In the nest box it provides a soft cushion so the egg doesn't crack.

There are several types of bedding material.

Straw

Although it compacts easily and isn't very absorbent it can be used for the coop floor but will need to be changed regularly. Because it's not very absorbent, it is more prone to mold. It is better used in the nest boxes than as floor litter.

Pine Shavings

Pine shavings, or any white wood shavings, is cheap, very absorbent and easy to use. It will act as insulation on the floor. This is good in the winter but in the summer make sure there is good ventilation to remove heat. You will need to have a plan for what to do with the soiled bedding.

Sand

In the coop, sand needs to only be replaced once or twice a year if regularly cleaned. It dries very quickly and can be turned over with a rake or scooped with a cat litter scooper.

For outdoor runs it doesn't break down, it dries fast and it doubles as a great material for dust baths.

Use construction sand, river sand, or builders sand. They're all the same product, just different places have a different name for it. You want the most course sand available.

If it starts to smell, sprinkle pelletized limestone in the sand not vermiculite.

To scoop the poop, cover a rake with hardware cloth to make a large litter scooper.

Materials not to use:

Hay gets damp and can cause fungal spores to grow.

Hardwood shavings have certain fungi and molds harmful to chickens.

Sawdust can cause respiratory problems.

Deep Litter Method

Let the bedding and poop compost on the coop floor instead of cleaning it out regularly. It helps to heat the coop, which in turn helps keep the chickens warmer. It should be stirred daily similar to turning a compost pile.

Extreme Heat

With the high heat in the summer, any bedding material you choose will dry out more quickly than during the rest of the year. Because of this we chose a sand floor we can wet down in the summer and not be concerned with creating mold or figuring out what to do with soiled bedding.

I don't recommend the Deep Litter Method in this part of the country because of the additional heat it creates.

Feed & Water

Adult chickens eat approximately 1/4 LB of food each day per bird. If you're not using an auto-feeder (meaning any style of bulk feeder), then feed the flock twice a day.

There are several types of feed on the market. So the first decision is which one is right for your flock.

Organic - grains grown organically, no pesticides or chemicals, non-gmo seeds with no cross contamination, no medication, no antibiotics, no hormones, protein must be organic. Non-organic ingredients allowed are vitamins, minerals, salt, and the amino acid methionine.

Conventional - grains and protein are from gmo sources, protein sources may have antibiotics, may contain corn and soybeans.

Medicated - this the same as conventional with medication added in to help chicks combat coccidiosis, usually contains the medication amprollium.

Chicks need to begin on Chick Starter from the time they hatch to 8 weeks. Chick Starter has 21% Protein which is needed to give them a good start growing. You can find Organic, Conventional and Medicated varieties at your local livestock store.

From 8 weeks onwards they can move to Grower/Layer Feed with 16-18% Protein. This comes in Organic or Conventional varieties.

Besides feed there are a few feed supplements that the flock will need.

Oyster Shell should be given starting at 16-18 weeks, in a separate feeder from the regular feed. The hens will take the oyster shell as needed to keep the calcium levels up for egg shell production.

Grit should be given starting at 8 weeks, in a separate feeder from the regular feed, when

Grower/Layer Feed is started. This is needed to help the crop grind up the feed. The chickens will take the grit as needed.

Scratch is a treat, very fattening and corn based. It should be used sparingly as a treat. It is thought that corn warms up the body from the inside. So not the best choice in the summer heat. More appropriate summer treats are covered in a later section.

Provide multiple sources of clean water.

1 TBS Apple Cider Vinegar (ACV) added per gallon of water will stop algea growth in the water container.

Galvanized containers can rust and leach zinc, especially if ACV is used in the water.

Extreme Heat

Combs, Waddles and Feet are used to disperse body heat, but chickens can't sweat.

Feed

In the summer especially, use 'no corn' feed since it's thought that corn warms up the body from the inside.

Water

Create Ice Blocks. To do this use any type of food storage container approximately 9Wx10Hx5D inch to make ice blocks for the adults and smaller 4Wx5Hx3D food storage container for any babies in the brooder or grow out pen. The container needs to have straight sides so you can pop the ice block out. Get a clay plate (the kind that goes under a clay planter), place a landscape brick on it and set the ice block on top. The clay and brick will hold the cold better than other materials. As the ice melts, the chickens will drink the cool water and stand in it (they cool down through their feet). If they are throwing too much bedding and dirt into the ice, raise the dish up off the ground to keep it clean, making sure the chickens have an easy way to climb up to it.

Misters

Some people setup misters in the shade, with or without a fan. Adding moisture to the dry hot air works similar to an evap system on a house. Just be sure not to create a mold issue if it's blowing over the bedding material and not blowing directly onto the chickens which may cause a respiratory issue. If the chickens feathers are getting wet from the mist, then the mister should be raised higher.

Frozen Treats

Cut melons or grapes in half, freeze them and serve frozen. The hens will peck at the frozen fruit. You can also freeze berries, cucumber pieces, bananas or peas. The frozen fruits can also be put in an ice block.

What not to use

Metal feed and water containers can get hot in the heat. The metal will conduct heat warming up the feed and water.

In feed containers, the heat can cause condensation dampening the feed and causing mold. If it's a treadle feeder the metal step could burn the bottom of their feet so look for a wooden one.

For water containers, a metal container could make the water too hot to drink at a time when preventing dehydration really matters so look for a plastic one.

Laying

There are many variables to hens laying eggs. The following can be causes for laying to slow down:

shorter days, less sunlight

extreme summer heat

stress in the flock

introduction of new members

When a hen is getting ready to start laying her comb and waddles will get very red, she will squat when you go to pet her back.

Kitchen scraps are ok but it will affect their protein and calcium ratio which might affect the laying frequency.

To tell if eggs are old, good, bad use the following float test:

sink - good egg

float - bad egg

stands on its end - old but not yet bad

Egg Shell Color

White ear lobes = white shell eggs

Red ear lobes = brown shell eggs

Blue ear lobes = blue shell eggs

Egg Storage

Store eggs large end up as that's where the air pocket is located.

Eggs have a natural clear coating on them called the Bloom. If they are washed, this bloom is removed and the eggs will need to be refrigerated.

If they are unwashed, they can be kept on the counter for a couple of weeks.

Extreme Heat

Collect eggs as often as possible through the day so they don't sit out in the extreme heat. If in question, use the float test.

Cleaning

When cleaning the coop use hot soapy water, then bleach water, then rinse with clean water.

Use a putty knife or hand hoe to scrape poo off all surfaces.

Protect any wood with natural oils.

Weekly Schedule

remove soiled litter

freshen nest boxes

Monthly Schedule

Freshen wet areas with SweetPDZ or Stall Dry

Freshen packed litter

Top dress with DE

Check for rodents

Annual Schedule

Clean all interior surfaces

Treat wood

Top dress surfaces for parasites

Freshen litter

Make any coop repairs

Broody

What is it?

A hen wishing to incubate eggs and raise chicks.

Symptoms?

She stays on her nest day and night, only leaving a few times per day to eat, drink, poop, and maybe dust bathe.

When you reach in to pet her or to remove eggs from under her, she will puff up her feathers to seem larger, get quite huffy yelling at you, or even pecking at you.

She may pluck out her own chest feathers to provide more warmth for the eggs directly from her bare skin.

After a few days of being broody she will stop laying and will not lay eggs again until she's no longer broody.

What to do?

Option #1

Give her fertile eggs to hatch or day old baby chicks to raise.

Option #2

Let her be and wait for the broodiness to pass, about 21 days. This isn't the best option because a broody hen will only take short breaks from the nest to care for herself. It can also induce other hens in the flock to go broody.

Option #3

If she isn't going to hatch eggs or raise new chicks, the best option is to break her from being broody.

Broody Breaking

Sometimes all it takes is a few times of being physically removed from the nest and carried out to the yard where her flock are.

Other times you will need to confine her to a wire-bottomed cage up on blocks so that air can flow up underneath her. The idea is to lower her body temperature and cool down her belly. Provide food and water, but NO bedding. Keep her in there for 3-4 days, unless she lays an egg earlier. Let her out one morning and watch what she does. If you find her back on the nest, she'll need a few more days in the Broody Breaker. But if she goes back to hang out with her flock, her mood has changed and broodiness is broken.

Misc

Clipping fly feathers

Fly feathers are the 10 longer primary wing feathers. They can be clipped to prevent the chicken from flying. You only need to clip one side and can start at 10 weeks.

Molting

Once a year, usually in autumn, chickens will shed their old feathers and grow new ones. This happens because new feathers are better winter insulators. Your coop and run might look like someone had a pillow fight and left the mess for you to clean up.

Chicks

The first molt is within the first 6 weeks or so, when the baby fluff is replaced with the first feathers.

Then there are several light or partial molts as these initial feathers are replaced with stronger feathers. This will happen around 7-9 weeks, 12-16 weeks, and again at 20-22 weeks. At this point the chicken will have all of its adult feathers.

Adults

The first adult molt usually happens around 18 months old.

No two chickens molt alike. Some molt quickly, others molt slowly. Some lose just a few feathers and others lose a lot.

It takes a lot of energy to grow new feathers. So during this process, your chickens might lose weight. Hens will stop laying, combs and waddles will pale.

Triggers

Chickens can molt at other times of the year. It can be triggered by extreme heat or big temperature changes, overcrowded flock, predator issues, illness, or other stressors.

How to Help

Feed them extra protein. This will help the new feathers come in. Some good choices are worms, black oil sunflower seeds, green beans, peas, fish and meats. There are several good molt muffin recipes on the internet that can be used as a treat.

Don't pick them up if they're having a hard molt. The pin feathers are sometimes uncomfortable when they come in.

Poop to Compost

Need to maintain a Carbon (C) to Nitrogen (N) ratio of 25:1. If the C:N ratio is too high, excess Carbon, decomposition slows down. 25:1 isn't the volume of brown material to green material, it's a chemical composition ratio.

Chicken poop is too strong to use as fertilizer right out of the coop. It needs to compost for a year.

Deep Litter Method

1st layer - dusting of sand

2nd layer - 4 inches of wood chips

3rd layer - 2 inches of straw or sawdust

Use pH strips to test acidity.

Keep the pile at 130 degrees for a week to kill any pathogens.

Turn the pile every week.

Good compost looks like coffee grounds and smells earthy.

By volume, use 1 part Brown to 2 parts Green mixture

Brown items (carbon rich used for energy by microbes)

leaves, straw, hay, sawdust, wood chips, twigs, newspaper, coffee filters, coop bedding, nuts, shells

Green items (nitrogen rich provides protein)

fruit and vegetable scraps, chicken poop, egg shells, tea bags, grass clippings, garden clippings, coffee grounds

Items not to use

meats, diary, high fat foods, peanut butter

Treats - Good/Bad

Bad

salt

dairy

chocolate, candy, sugar

raw potato or green peels

nothing moldy/slimy that has gone bad

citrus

dried or undercooked Beans

avocado skin and pit

Good

Apples: raw or applesauce but not the seeds

Asparagus: raw or cooked

Bananas: no peel

Beans: well-cooked only, not dry

Beets: flesh and greens

Berries: all kinds

Breads: all kinds but in moderation, stale bread is ok

Broccoli & Cauliflower

Cabbage & Brussels Sprouts

Carrots: raw and cooked, flesh and greens

Cheese including cottage cheese: feed in moderation

Cooked Chicken

Corn: on the cob, canned, raw and cooked

Crickets

Cucumbers

Eggs: hardcooked and scrambled

Eggplant

Fish / Seafood: cooked only

Flowers: marigolds, nasturtiums, pansies, etc. not treated with pesticide

Fruit: pears, peaches, cherries, apples, grapes, melons but no seeds or pits

Grains: bulgar, flax, niger, wheatberries, etc.

Green beans

Grits: cooked

"Leftovers": Only feed your chickens that which is still considered edible by humans, don't feed anything spoiled, moldy, oily, salty or unidentifiable.

Leafy Greens: lettuce, kale, spinach collards, chickweed, chard, etc

Mealworms

Oatmeal: raw or cooked

Pasta / Macaroni: cooked spaghetti, etc.

Peas: peas, pea tendrils and flowers

Peppers (bell)

Pomegranates: raw

Popcorn: popped, no butter, no salt

Potatos / Sweet Potatos/Yams: cooked only, avoid green parts of peels

Pumpkins / Winter Squash: raw or cooked

Raisins

Rice: cooked only

Scratch

Sprouts

Summer Squash: yellow squash and zucchini

Sunflower Seeds

Tomatos: raw and cooked, no green tomatoes or plant stems/vines

Turnips: cooked

Yogurt: plain

Predators

Possible Predators by Condition

If adult birds are missing but no other signs of disturbance exist.

It's probably a Dog, Bobcat, Coyote, Fox, Raccoon, Hawk, or Owl. These predators typically are able to kill, pick up, and carry off an adult chicken. Hawks typically take chickens during the day, whereas owls take them during the night.

If chicks are missing but no other signs of disturbance exist.

It's probably a Snake, Rat, Raccoon, or house cat. Such predators sometimes leave some feathers and wings scattered away from the site because they are not able to swallow these parts.

If birds are dead but not eaten and have parts still intact.

A weasel may have attacked the flock. Members of the weasel family, including mink, kill just for the fun of killing. Often, the chickens' bodies are bloodied. Also, you might notice that internal organs have been eaten.

If birds are dead and not eaten but are missing their heads,.

The predator may be a raccoon, a hawk, or an owl. Raccoons sometimes pull a bird's head through the wires of an enclosure and then can eat only the head, leaving the majority of the body behind. Also, raccoons may work together, with one scaring the chickens to the far end of a pen and the other picking off the birds' heads.

If birds are only wounded, not dead

Various predators may be to blame. If birds show signs of bites all over, a dog may have attacked the flock. Dogs do not have sharp enough teeth to consume animals cleanly. If the wounds are on the breasts or legs of young birds, an opossum may be the problem. Bites on the hocks of young birds often indicate that rats have preyed on the flock. If birds have bites and show signs that their intestines have been removed through their cloacae, the attacker may be a member of the weasel family, or cannibalism may be occurring in the flock.

If eggs are missing

One of several predators—including skunks, snakes, rats, opossums, raccoons, blue jays, and crows—may be at fault.

Predator Behaviors

Domestic Dog

Domestic dogs that run free often kill for the fun of it. Not all dogs will attack, it depends on breed as some have a greater tendency to chase prey.

Coyote

Usually hunt in pairs. They are primarily active at night but have been seen during the day.

Bobcat

They prefer to hunt at twilight and dusk, but will take prey anytime. The bobcat will carry the chicken off.

Fox

Usually attack the throat but some will use multiple bites to the back and neck. The fox will carry the chicken off and also eat eggs.

Raccoon

They will takes several birds in one night. They prefer the crop and entrails. They will also eat eggs.

Weasel

The Least Weasel is the smallest predator, weighing 1-2 oz and only 6-8 inches long so they can fit through 1/4 inch hardware cloth. They hunt day and night.

Snake

They are hard to identify because they eat their prey whole so there's no signs other than a missing egg or chick. The size of the hole needed depends on the size of the snake. But the hole also needs to be large enough that the snake fits after eating the prey. Usually snakes that fit through 1/4 inch hardware cloth don't pose a threat.

Hawk

The most common hawks that prey on poultry are red-tailed, hawk (sometimes called a chicken hawk) red-shouldered hawk, and cooper's hawks.

They hunt during the day and usually carry it off leaving no signs other than a missing bird.

Owl

The most common owl that preys on poultry is the great horned owl. Owls are more active at night, and that is when they typically take birds.

Prevention

When free ranging

Use moveable fences. They may be electrified to stun. Eliminate any perch sites within 100 yards of the flock.

Get guard dogs that are trained to help manage a flock.

Net Covered Run

Use orange colored netting as hawks and owls can see orange well. Make sure it's secure so they can't get through any weak spots.

Permanent Run

Cover the run in welded wire. Make sure the predators can't dig their way into the run by burying hardware cloth 2 inches below the surface and out another 18 inches from the wall.

Ailments, Diseases, Treatment

Disclaimer: I am not a veterinarian or trained medical practitioner. This information is based on my own research and experience with my flock. Please be diligent in doing your own research to determine the best course of action for your flock. All remedies don't work the same on all birds. What might work for one, may not work for another.

Preventative

At 18 weeks you can administer worm control.

Bumblefoot

What is it?

A bacterial infection and inflammatory reaction on the feet creating a pus filled abscess. This is caused by Staphylococcus bacteria referred to as "plantar pododermatitis" by medical professionals. It can be fatal if left untreated because the infection can spread to other tissues and eventually to the bones.

Symptoms?

Indications are swelling, sometimes redness, and often a black or brown scab on the bottom of the foot.

Treatment?

Natural #1

Soak in warm water with Epsom Salt for 20 - 30 minutes to soften the scab

Spray with Vetericyn then apply Green Goo or drawing salve

Wrap in gauze and secure with vet wrap and first aid tape

Repeat twice a day

If not healing, then surgery is needed to cut out the kernel

Natural #2

Apply Bumblefoot Remedy by Moonlight Mile Herb Farm as directed in the package.

Surgery

Generally takes about 1 hour to complete. It can be done by one person, two makes it much easier.

Chickens should not be given anesthetics, such as benzocaine or xylocaine, which are known to cause heart failure and death. Even over-the-counter products, such as triple-antibiotic ointment containing topical anesthetics ending in the suffix "caine," can be fatal. You can request your veterinarian to prescribed Metacam, a pain-relieving medication for pets, which is safe for use in chickens, to keep on hand.

Ideally, a chicken with bumblefoot should be treated by an avian veterinarian. If you don't have access to one, there are several websites that show how to do this procedure at home.

Prevention

Regularly inspect your chickens' feet to detect any instances of infection at the earliest possible stage.

Prevent obesity and avoid vitamin deficiencies, both of which are a risk for contracting bumblefoot.

Keep roosts splinter-free and less than 18 inches from the floor, so that chickens don't hurt themselves with repetitive jumping.

Ensure coop litter is clean and dry at all times resulting in cleaner chicken feet.

Mites and Poultry Lice

What is it?

Mites and Poultry Lice are a part of every backyard, so when your flock becomes infested it doesn't mean you're not keeping a clean coop.

Poultry Lice are not the same as human head lice and people can't contract lice from chickens.

Most common Mites are Northern Fowl Mites and Red Roost Mites. They live both on the chicken and in the coop. They are active at night to leech blood from the chicken while they sleep.

Poultry Lice only live on the chicken and its feathers. They feed on dead skin cells and feather quill casings.

Scaly Leg Mites are microscopic insects that live under the scales on the chickens legs and feet. They dig tunnels, eating the tissue, leaving their crusty waste.

Symptoms?

Mites and Poultry Lice

Inspect your flock monthly for dirty looking vent feathers, a pale comb, feather pulling, bald spots, redness or scabs on the skin, dull ragged looking feathers and bugs or nits. You might also notice changes in appetite, drop in egg production, weight loss, decreased activity or listlessness.

Scaly Leg Mites

Thick, scabby, crusty looking areas on legs or feet.

If one bird is infected, all should be treated.

Treatment?

Mites and Poultry Lice

Some treatments are Garden and Poultry Dust with Permethrin by YTex, Poultry Protector by Manna Pro, and Elector PSP by Elanco (no egg withdrawal).

Garden and Poultry Dust is easier to apply at night when the birds go to roost. Dust under the wings and vent area thoroughly.

Elector PSP by Elanco is recommended by poultry veterinarians. Spray the cleaned coop, and spray the birds with a mixture of 9ml Elector per gallon water.

Most applications need to be repeated every 7 days until all nits have hatched and been treated.

Scaly Leg Mites

Soak the feet and legs in warm water to loosen the scales. Dry them off with a towel, gently exfoliating any older scales. Dip the legs and feet in oil (olive oil, vegetable oil, whatever you have available in the kitchen) to suffocate the mites. Wipe off the excess oil and slather the affected area with petroleum jelly. Reapply the petroleum jelly daily until the scales are back to normal.

Additional information can be found on the Mississippi State University Extension Service website.

Preventative?

Chickens dust bathe to keep parasites away. Sand and wood ash are sufficient.

Avian Pox/Fowl Pox

What is it?

A viral disease contracted by mosquitoes, other chickens with pox, or contaminated surfaces. Recovered birds do not remain carriers.

Symptoms?

White spots on the skin; combs turn into scabby sores; white membrane and ulcers in mouth, on trachea; laying stops; all ages affected.

Treatment?

Supportive care, warm dry quarters, soft food; many birds with good care will survive.

Vaccine available: Yes

Impacted Crop

What is it?

A blockage in the crop. This is usually caused by tough strands of grass or other things that get stuck that cannot pass through the crop.

Symptoms?

The crop is full and feels like dough.

It is normal for the crop to look big when full. So check the crop in the morning before a bird has eaten to determine if it emptied naturally.

Sour Crop is a similar condition which is caused by a yeast in the crop and recognized by a putrid smell coming from the beak. Treatment is the same, requiring emptying the crop.

Treatment?

Mild case:

Epsom Salt detoxifies toxins if your chickens get into something toxic. It acts as a laxative, flushing the system if the crop is impacted or the digestive process seems blocked. This will gently move things along. 1 teaspoon of Epsom salt in 1 ounce of water ONLY ONE TIME TREATMENT! Feed orally with a syringe or isolate her with only the Epsom Salt water to drink. She will be extremely thirsty so be sure to keep lots of water available!

Sever case:

The contents of the crop can be softened with a couple of teaspoons of Olive Oil in warm water poured down the throat. Massage the crop for 5 minutes. Then turn the bird upside down and massaging the crop for a few seconds at a time to empty the crop. There is a risk of the bird choking so you must allow the bird a chance to catch its breath in between. This doesn't always work (you can't always completely empty the crop) so if you are in any way unsure, it is better to allow a vet to perform a procedure to cut the crop open and empty it under a local anaesthetic.

Vent Gleet

What is it?

An avian yeast infection also known as cloacitis, thrush, or mycosis. It is a fungal Infection.

Symptoms?

Dirty feathers with poo, mucus around the vent, and sometimes nasty smelling discharge. The infection may also cause a red and/or swollen vent which may bleed. Scarring may result with associated reduction in the elasticity and diameter of the vent leading to problems with egg laying and defecation. Other general signs of illness often include fluffed feathers, a hunched appearance, partially closed eyes, the head tucked under a wing, sitting or standing on the ground rather than roosting.

Treatment?

Soak and wash the vent area with Epson Salt water for 15 - 20 minutes. Use a little Dawn if the poo won't soften. You can even cut away the feathers if it's really bad.

Add 1.5 TBS ACV to 2 cups of drinking water as vinegar prevents yeast growth. Administer a few TBS of yogurt or other pro-biotic each day.

Even if you see improvement, maintain the treatment for 7 days to prevent relapse.

Isolate the infected bird to prevent spread of any secondary infections and to avoid other birds pecking at the red/bloody vent.

If more than 1 bird is infected, have some feces examined by your Vet to identify the type of cloacal infection bacteria, parasite, fungus or yeast.

Infectious Bronchitis:

What is it?

Viral disease; highly contagious; spreads through air, contact, and contaminated surfaces.

Symptoms?

Coughing; sneezing; watery discharge from nose and eyes; hens stop laying.

Treatment?

Supportive care; 50 percent mortality in chicks under 6 weeks.
Vaccine available: Yes. Give to hens before 15 weeks of age because vaccination will cause laying to stop.

Infectious Coryza

What is it?

Bacterial disease; transmitted through carrier birds, contaminated surfaces, and drinking water.

Symptoms?

Swollen heads, combs, and wattles; eyes swollen shut; sticky discharge from nose and eyes; moist area under wings; laying stops.

Treatment?

Birds should be destroyed as they remain carriers for life.
Vaccine available: None.

Mareks Disease

What is it?

Viral disease; very contagious; contracted by inhaling shed skin cells or feather dust from other infected birds.

Symptoms?

Affects birds under 20 weeks primarily; causes tumors externally and internally; paralysis; iris of eye turns gray, doesn't react to light

Treatment?

None; high death rate and any survivors are carriers.
Vaccine available: Yes, given to day old chicks.

Mycoplasmosis/CRD/Air Sac Disease

What is it?

Mycoplasma disease; contracted through other birds (wild birds carry it); can transmit through egg to chick from infected hen.

Symptoms?

Mild form — weakness and poor laying.

Acute form — breathing problems, coughing, sneezing, swollen infected joints, death
How contracted: Mycoplasma disease; contracted through other birds (wild birds carry it); can transmit through egg to chick from infected hen.

Treatment?

Antibiotics may save birds — see a vet.
Vaccine available: Yes.

Dehydration

What is it?

The effects of heat on chickens is cumulative. It's important to keep the flock hydrated in the extreme heat. Chickens don't sweat. Instead they hold their wings away from their body to increase air flow. They will pant, increasing their respiratory and heart rate which can lead to Acidosis. Their combs, waddles, and feet act as radiators letting heat escape their body.

Chickens can suffer from Heat Stress or Dehydration when temperatures increase suddenly or rise above 85 degrees. Sudden changes in temperature is more dangerous than a gradual change. The added blood flow to the combs and waddles to attempt to cool down reduces the flow to their organs.

Heat stress and dehydration deplete the body of electrolytes required for a chicken's normal body functioning, therefore replenishing them is a priority when chickens suffer from heat stress and/or dehydration.

ACV should NOT be added to waterers during times of high heat. During high heat conditions, baking soda facilitates calcium absorption while ACV inhibits it.

Symptoms?

Before dehydration sets in, birds may pant, open their wings, and fluff out their feathers. These are the first warning signs.

The first symptom that tends to turn up is paleness of the face. Breathing becomes heavy and labored. After a while the chicken will develop diarrhea. If you gently pinch the back of the shank (lower leg) the skin will not spring back as usual, much like the test performed on dehydrated humans. Shortly after the diarrhea, the bird will become listless, limp or even completely unreactive. If the dehydration goes too far the bird will go into convulsions. These involve unconscious twitching of the muscles, backward arching of the neck and paddling of the feet. This is broken by periods of general limpness and unreactivity. Shortly after, unless immediate action is taken, comes death.

Treatment?

Place the bird in a cool, dark place to keep it calm. Wrap it in a towel if you need to syringe feed and don't have someone to help you hold the bird.

Provide water with electrolytes. Dip the bird's beak into the water a couple of times and, if need be, help it tilt it's head back to swallow. Wait 5-10 minutes then repeat for the next hour or so. Increase time between waterings. Once the bird drinks on it's own you can put out feed moistened with water. Leave the chicken in a cool environment with plenty of water and feed for about 24 hours to ensure recovery.

There are several ways to get electrolytes to add to drinking water:

Make a home recipe of electrolytes

Purchase tablets or powder at any livestock or feed store

Purchase the unflavored Pedialyte at a local grocery store

To entice the flock to drink more, make frozen treats with or without fruit, but add the electrolytes. See more details in the Feed & Water section.

Scorpions

What is it?

Most scorpions are nocturnal hunters and chickens are usually off the ground roosting somewhere all night. Scorpion stings contain a neurotoxin, it usually makes the chicken act like it's had a stroke for a couple of days.

If a chicken grabs a scorpion in the wrong way and gets stung on the face or mouth it can be a rough situation, though they usually pull through. The greatest danger is that their throat swells up and that kills them.

Symptoms?

Swelling in the face, general physical weakness, little or no appetite.

Treatment?

Keep her still and secluded inside where it's cool. Crush one 25 mg Benadryl tablet in 2 T. of warm water. Mix until dissolved, then using a syringe (the kind that one gives children liquid medication with) give the chicken 2 mL of the suspension.

In 72 hours she should be back to normal.

Unexpected Losses

Sometimes no matter how diligent and careful you are, things just go wrong. Even with the best care a chicken can get sick, have an accident, or a genetic issue you were't aware of. It can be hard to find a vet who will see a chicken and even when you find one there are only a few medications approved for use on chickens. You can only do your best!

Essential Oils and Herbs

Disclaimer: I am not a veterinarian. Please use common sense and do your research before using alternative remedies with your flock. When in doubt, contact an avian veterinarian before treating your birds.

Using essential oils and chicken care can be a wonderful combination in a more natural approach to chicken keeping. As with anything that you are doing to help improve your farming practices, make sure to do your own research and ask questions until you are comfortable that the choice is the right one for you to use.

When you are using essential oils around your animals, always make sure you are using diluted strengths. Some oils, such as peppermint or oregano are considered "hot" oils and can burn skin. Always research the oil and its uses before using on any animal and make sure that you dilute the oil with an oil such as fractionated (liquid) coconut oil or liquid almond oil before applying.

Citrus Essential Oils for Cleaning

Lemon and other citrus fruits have natural cleaning and disinfecting capabilities. You can easily make a homemade coop cleaner. Spray down the coop, allow the damp areas to dry completely before adding clean bedding. You can also use this to clean and disinfect feed bowls, water founts, cages, and the chicken crate used as a hospital for your sick or injured birds. Rinse well before re-using.

8 oz white vinegar

8 oz water

25 drops of any citrus essential oil

Wound Care Ointment

Frankincense, Lavender and Coconut Oil, promotes healing for superficial wounds, cuts and scrapes. Makes a solid paste that can be scooped out with your fingers to apply to the wound. The ointment softens quickly, and smells pleasant.

This wound ointment is the one I use for superficial wounds, cuts and scrapes.

4 ounce container with a lid

4 ounces of solid coconut oil

12 drops lavender essential oil

12 drops frankincense essential oil

Melt the coconut oil, add the essential oils and mix. Allow to harden in the container. Ready for use! If you leave it in a warm area it will liquify. To prevent this you can also add melted bees wax to the recipe for a more solid ointment.

Oregano as a Feed Additive for Internal Health

Oregano essential oil has many types of uses from wound care to intestinal worms, to a replacement for routine antibiotics in flock health care and prevention. Bell and Evans Poultry Farm in Pennsylvania switched to using an oregano oil based feed additive instead of using any antibiotics and had great results.

Clove Oil

Clove oil diluted at 2% in a carrier oil (such as olive) can be used to prevent feather picking. This works well if your bird is picking her own feathers or if others are picking at her. Just remember, if a hen is picking her own feathers, there is most likely an underlying cause. Treat the problem before dealing with the side effect.

Lavender Oil

When made into a salve with coconut oil, lavender oil is great at healing and protecting open wounds and sore. Dilute to 1% before applying and be sure to avoid the eyes, mouth, and nostrils.

Neem Oil

Neem oil disturbs the life cycle of scaly mites. Apply at a 1-2% dilution with soybean oil, linseed oil, or vegetable oil on affected chickens. Follow up with a thick coating of Vaseline.

HERBS

When using herbs inside the coop, make sure they are first thoroughly dried. Moist and fresh herbs mixed in with organic bedding may cause molding.

Insect Repellant

They can be grown around your coop or dried and hung inside.

Catnip

Fennel

Feverfew

Lavender

Pennyroyal

Peppermint

Rosemary

Spearmint

Tansy

Rodent Repellant

Use them like you would the insect-repelling herbs, or make satchels to store in grain bins and bags.

Lemon Balm

Mint

Nest boxes

Relaxants

Dandelion, dill, lavender, lemon balm, and rose hips

Increase Egg Production

Borage, comfrey, fennel, marigold, marjoram, mint, nasturtium, parsley, sage, and thyme

Ward off internal parasites

Hyssop, nasturtium, sage, spearmint, tansy, and thyme

OTHER NATURAL REMEDIES

Diarrhea? Try offering wheat bran soaked in buttermilk.

Sour crop? Withhold food and water for 24 hours. Give a tablespoon of coconut oil or olive oil via an eyedropper. Wait 12 hours and then offer scrambled egg mixed with plain yogurt.

Wry neck? Try Turmeric Tea in the same dose you would drink yourself.

First Aid Kit

Rooster Booster No Pick

Apple Cider Vinegar

Rubber Gloves

Neosporin - not triple action (no caines)

Antibiotic

Vetericyn - kills bacteria, virus, fungi, ringworm

Nutri Drench - boost immune system, liquid vitamin

Kocci Free - anti-parasitic

Poultry VetRx - antibiotic, respiratory, scaly leg, eye worm

VermX - worm preventative

Garlic Juice - mites & lice

Honey - antiseptic healing

Saline Solution - cleaning wounds or eyes

Cornstarch - stop bleeding

Molasses - cause diarrhea to flush the system

Oregano Oil - antibiotic

Waxelene

Liquid Childrens Benadryl - stings

Liquid Calcium - egg bound

Electrolytes or Pedialyte - dehydration

Epsom Salts - bumble foot, splinter, vent gleet

Blue Kote - antiseptic, anti-pick

Oxine AH - disinfectant

Gauze Pads

First Aid Tape

Vet Wrap

Wooden Popsicle Sticks

QTips

Scalpel

Eye Dropper

Tweezers

Small Pliers

Plastic Syringe

Dog Toenail Clipper

Small Flashlight

Chicken Saddle

Pet Carrier

Blanket

Resources

AZ Dept of Agriculture

(602) 542-4373

AZ Veterinary Diagnostic Lab

(520) 621-2355

acts.cals.arizona.edu

State Animal Official

Dr. Perry Durham

1688 W. Adams St, Phoenix AZ 85007

(602) 542-4293 or (602) 542-4290

pdurham@azda.gov

AZ Exotic Animal Hospital

20040 N. 19th Avenue Suite C, Phoenix AZ 85027

(623) 243-5200

Scottsdale Livestock

22255 N. Scottsdale Rd, Scottsdale AZ

(480) 515-1800

Western Ranchman

16028 N 32nd St, Phoenix AZ

(602) 992-3410

Facebook Groups / Blogs

Fresh Eggs Daily

www.fresheggsdaily.com

The Egg & I

www.the-eggs-and-i.com

Timber Creek Farmer

www.timbercreekfarmer.com

104 Homestead

www.104homestead.com

Instructables

www.instructables.com

Mississippi State University Extension Service

http://extension.msstate.edu/

Pennsylvania State University - Dr Eva Wallner-Pendleton

http://extension.psu.edu/animals/poultry/courses/pennsylvania-poultry-sales-and-service-conference-and-northeastern-conference-on-avian-diseases/presentations/2014/layer-program/external-internal-parasites-of-poultry

Cooperative Extension System

extension.org

Maricopa County Food System Coalition

http://forum.vpaaz.org/forum/topics/chickens-and-scorpions-1

Poultry Keeper

poultrykeeper.com

Bell and Evans Poultry Farm

http://www.bellandevans.com/

The Prairie Homestead

http://www.theprairiehomestead.com/how-to-purchase-essential-oils

Books to keep for Reference

Fresh Eggs Daily by Lisa Steele

Chicken Health Handbook by Gail Damerow

Fowl Play by Rachel Bess

www.ingramcontent.com/pod-product-compliance
Lightning Source LLC
Chambersburg PA
CBHW070202230526
45471CB00002B/787